Keto Meals For Beginners

Scrumptious Keto Meals For Weight Loss And Revitalize Your Body

Celine King

TABLE OF CONTENT

Introduction

The ketogenic diet is a very low carb, high-fat, and moderate protein diet. The idea behind the ketogenic diet is that the liver will break down fatty acids into ketones, which can fuel the brain. Ketogenic diets are often prescribed for children with intractable epilepsy because they may experience seizure control while on this type of eating plan. There are many reasons to try a ketogenic diet, but the biggest reason is weight loss. This diet allows you to lose weight without training a lot, as your body burns fat for energy instead of carbohydrates. As a result, you can have more energy, feel better, and have fewer cravings.

An essential step in any diet is considering and establishing goals–thinking hard about being able to precisely articulate what you wish to achieve by getting started on a diet. Because of the temptations of food and treats that can cause us to watch out for health-wise, a keto diet is a fantastic option due to all of the ailments it can help prevent or supplement.

Most people can safely seek out the keto diet. Nonetheless, it is best to talk to a dietitian or doctor about any significant diet changes. It is the case for those with disabilities underlying it.

A successful treatment for people with drug-resistant epilepsy could be the keto diet.

While the diet can be ideal for people of any age, children, and people over the age of 50, infants may enjoy the most significant benefits as they can easily adhere to the diet.

Adolescents and adults, such as the modified Atkins diet or the low-glycemic index diet, can do better on a modified keto diet.

A health care provider should track closely; whoever is using a keto diet as a medication. A doctor and dietitian can monitor a person's progress, prescribe medications, and test for adverse effects.

The body processes fat differently from that it processes protein differently from that of carbohydrates. The Carbohydrate response to insulin is extreme. The protein response to insulin is moderate, and the fast response to the insulin is minimal. Insulin is the hormone that produces fat / conserves fat.

After you've planned out your protein and carbohydrates, eat fat. You can eat all the fat you want as long as you're not doing it to excess. But unlike Weight Watchers or other diet plans, you don't need to measure fat or count calories. Simply let your body tell you when you've had enough. If you eat fat until you're satiated, you won't have problems consuming too many calories. If you eat and still feel like you need to eat more – do it. Many beginners on keto get into trouble when they don't eat enough fat. Fasting can be incorporated into the keto diet if it's done correctly. Try out one of the intermittent fasting techniques to help accelerate and maintain weight loss after you're adapted to keto.

Electronic monitors can be beneficial to keep track of your progress at home. If you can afford it, you should get a blood sugar monitor and a ketone monitor. Track your fasting blood sugars and keep track of your ketones, ensuring they fall within the 1.5-3.0 mmol/dL range. Also, you may want to track your HDL and triglycerides. Home monitors can be used to do this and allow you to monitor progress more frequently and keep away from unnecessary trips to the doctor's office.

Lastly, remember to keep a journal. It's essential to keep track of your progress and helps you note not only how your triglycerides may be improving, but if you write down what you eat and find out you're not losing weight, it will make it easier to pinpoint problem areas where you can improve.

BREAKFAST

Kale Wrapped Eggs

Preparation Time: 8-10 minutes

Cooking Time: 5 minutes

Servings: 4

Ingredients:

- Three tablespoons heavy cream
- Four hardboiled eggs
- ¼ teaspoon pepper
- Four kale leaves
- Four prosciutto slices
- ¼ teaspoon salt
- 1 ½ cups water

Directions:

1. Peel the eggs and wrap each with the kale. Wrap them in the prosciutto slices and sprinkle with ground black pepper and salt. Place it in your instant pot and pour water. Arrange the eggs over the trivet/basket.
2. Close the lid and lock it. Press "MANUAL" cooking function; timer to 5 minutes with default "HIGH" pressure mode. Allow the pressure to build to cook. After cooking time is over press "CANCEL" setting. Find and press "QPR" cooking function. This setting is for quick release of inside pressure.
3. Slowly open and take it out from the lid.

Nutrition:

247 Calories

20g Fat

19g Protein

Zucchini Keto Bread

Preparation Time: 8-10 minutes

Cooking Time: 40 minutes

Servings: 12-16 slices

Ingredients:

- 1 cup grated zucchini
- 2 ½ cups almond flour
- ½ cup chopped walnuts
- 3 eggs
- ½ cup olive oil
- 1 ½ teaspoon baking powder
- Pinch of ginger powder
- 1 teaspoon vanilla extract
- ½ teaspoon cinnamon
- ¼ teaspoon nutmeg
- pinch of sea salt
- 1 ½ cups water

Directions:

1. Whisk together the wet ingredients in a bowl. Combine the dry ingredients in another bowl. Combine the dry and wet mixture. Stir in the zucchini.
2. Grease a loaf pan and pour the mixture. Top with chopped walnuts. Open its top lid and pour water. Arrange a trivet or steamer basket inside that came with Instant Pot. Now arrange the loaf pan over the trivet/basket.
3. Close the lid and press "MANUAL" cooking function; timer to 40 minutes with default "HIGH" pressure mode. Allow the pressure to build to cook the ingredients. After cooking time is over press "CANCEL" setting. Find and press "QPR" cooking function. This setting is for quick release of inside pressure.
4. Slowly open the lid, take out the cooked bread. Cool down, slice, and serve.

Nutrition:

164 Calories

17g Fat

5g Protein

Ham Sausage Quiche

Preparation Time: 8-10 minutes

Cooking Time: 30 minutes

Servings: 4

Ingredients:

- 4 bacon slices, cooked and crumbled
- ½ cup diced ham
- 2 green onions, chopped
- ½ cup full-fat milk
- Six eggs, beaten
- 1 cup ground sausage, cooked
- 1 cup shredded cheddar cheese
- ¼ teaspoon salt
- Pinch of pepper
- 1 ½ cups water

Directions:

1. Grease a baking dish with coconut oil cooking spray. Place all of the ingredients in a bowl, and stir to combine. Add this mixture to the prepared dish.
2. Open its top lid and pour water. Arrange a trivet or steamer basket inside that came with Instant Pot. Now arrange the dish over the trivet/basket.
3. Close the lid and press "MANUAL" cooking function; timer to 30 minutes with default "HIGH" pressure mode. Allow the pressure to build to cook. After cooking time is over press "CANCEL" setting. Find and press "QPR" cooking function. This setting is for quick release of inside pressure.
4. Place the dish on the rack in your IP and close the lid. Cook on high and release the pressure naturally. Slowly open the lid, take out the cooked recipe in serving plates or serving bowls, and enjoy the keto recipe.

Nutrition:

398 Calories

31g Fat

26g Protein

Coconut Almond Breakfast

Preparation Time: 8-10 minutes

Cooking Time: 5 minutes

Servings: 2

Ingredients:

- 2 tablespoons roasted pepitas
- 1/3 cup coconut milk
- 2 tablespoon chopped almonds
- 1 tablespoon chia seeds
- 1/3 cup water
- One handful blueberries

Directions:

1. Mix the pepitas with almonds and blend well. Switch on the pot. Add the chia seeds with water and coconut milk; gently stir to mix well. Add the pepita mix and combine.
2. Close the lid and press "MANUAL" cooking function; timer to 5 minutes with default "HIGH" pressure mode. Allow the pressure to cook.
3. After cooking time is over press "CANCEL" setting. Find and press "QPR" cooking function. This setting is for quick release of inside pressure. Slowly open and take out the dish from the lid.

Nutrition:

148 Calories

6g Fat

2g Protein

Avocado Egg Muffins

Preparation Time: 8-10 minutes

Cooking Time: 12 minutes

Servings: 4

Ingredients:

- 1 ½ cups of coconut milk
- 2 avocados, diced
- 4 ½ ounces (grated or shredded) cheese
- ½ cup almond flour
- 5 bacon slices, cooked and crumbled
- 5 eggs, beaten
- 2 tablespoon butter
- 3 spring onions, diced
- 1 teaspoon oregano
- ¼ cup flaxseed meal
- 1 ½ tablespoon lemon juice
- 1 teaspoon minced garlic
- 1 teaspoon onion powder
- 1 teaspoon salt
- Pinch of pepper
- 1 teaspoon baking powder
- 1 ½ cups water

Directions:

1. Whisk together the wet ingredients. Stir in the dry ingredients gradually until turns smooth. Stir in the avocado, bacon, onions, and cheese. Add the mixture into 16 muffin cups. Arrange Instant Pot over a dry platform in your kitchen. Open its top lid and switch it on.
2. In the pot, pour water. Arrange the 8 cups over the trivet/basket.
3. Close the lid and press "MANUAL" cooking function; timer to 12 minutes with default "HIGH" pressure mode. Allow the pressure to cook.
4. After cooking time is over press "CANCEL" setting. Find and press "QPR" cooking function. This setting is for quick release of inside pressure. Slowly open and take out the dish from the pot.
5. Repeat the same process.

Nutrition:

146 Calories 11g Fat

6g Protein

Kitchen Corner - Try It

Soft-Boiled Eggs

Preparation Time: 5 minutes

Cooking Time: 3 minutes Servings: 4

Ingredients:

- 4 Eggs
- 2 cups Water

Directions:

1. Switch on the instant pot, pour in water, insert steamer basket and place eggs in it.
2. Shut the instant pot with its lid in the sealed position, then press the 'manual' button, press '+/-' to set the cooking time to 3 minutes and cook at low-pressure setting; when the pressure builds in the pot, the cooking timer will start.
3. When the instant pot buzzes, press the 'keep warm' button, do a quick pressure release and open the lid.
4. Fill a bowl with ice water, place eggs in it from the instant pot, and let rest for 3 minutes.
5. Then peel the eggs, cut into slices, season with salt and black pepper and serve.

Nutrition:

68 Calories

4.6g Fat

5.5g Protein

Breakfast Casserole

Preparation Time: 10 minutes

Cooking Time: 45 minutes

Servings: 6

Ingredients:

- 1/2 teaspoon salt
- 2 tablespoons avocado oil
- 6 ounces breakfast sausage
- 1 1/2 cups Broccoli stalks, grated
- 1 tablespoon Minced garlic
- ½ teaspoon Ground black pepper
- 6 eggs
- 1/4 cup Heavy cream
- 1 cup Monterey jack cheese, grated
- 1 cup Water
- 1 Green onion sliced
- 1 California avocado, sliced
- ¼ cup Sour cream

Directions:

1. Switch on the instant pot, grease the pot with oil, press the 'sauté/simmer' button, and add the sausage and cook until the meat is no longer pink.
2. Then add broccoli along with garlic, season with salt and black pepper and continue cooking for 2 minutes.
3. Take a 7-inch baking dish, grease it with oil, spoon in cooked broccoli mixture and spread evenly.
4. Crack the eggs in a bowl, add cream, whisk until combined, and then add onion and cheese, whisk until mixed, pour the mixture over the sausage mixture and cover with aluminum foil.
5. Press the 'keep warm' button, wipe the instant pot clean, pour in water, then insert trivet stand and place baking dish on it.
6. Shut the instant pot with its lid in the sealed position, then press the 'manual' button, press '+/-' to set the cooking time to 35 minutes and cook at high-pressure setting; when the pressure builds in the pot, the cooking timer will start.
7. When the instant pot buzzes, press the 'keep warm' button, release pressure naturally for 10 minutes, then do quick pressure release and open the lid.
8. Take out the baking dish, uncover it and turn it over the plate to take out the frittata.
9. Top the frittata with avocado, cut into slices and the top with sour cream.

Nutrition:

351 Calories

28.5g Fat

18.6g Protein

Poblano Cheese Frittata

Preparation Time: 5 minutes

Cooking Time: 35 minutes Servings: 4

Ingredients:

- 4 Eggs
- 10 oz. Diced green chili
- 1 tsp. Salt
- ½ tsp. Ground cumin
- 1 cup Mexican cheese blend, shredded, divided
- ¼ cup Chopped cilantro
- 2 cups Water

Directions:

1. Crack eggs in a bowl, add green chilies, half-and-half, and ½ cup cheese, season with salt and cumin, stir well until incorporated. Take a 6-inch baking dish or silicone pan, grease it with oil, pour in the egg mixture and cover with aluminum foil.
2. Switch on the instant pot, pour water in it, then insert trivet stand and place baking dish on it. Shut the instant pot with its lid in the sealed position, then press the 'manual' button, press '+/-' to set the cooking time to 20 minutes and cook at high-pressure setting; when the pressure builds in the pot, the cooking timer will start.
3. When the instant pot buzzes, press the 'keep warm' button, release pressure naturally for 10 minutes, then do a quick pressure release and open the lid. Meanwhile, switch on the broiler and let it preheat.
4. Take out the baking dish, spread remaining cheese on top, then place it under the broiler and broil for 5 minutes or until cheese melts and the top is nicely browned.
5. When done, turn the dish over a plate to take out the frittata, then cut into slices and serve.

Nutrition:

257 Calories 19g Fat 14g Protein

Poached Egg

Preparation Time: 5 minutes

Cooking Time: 7 minutes Servings: 4

Ingredients:

- ¾ teaspoon salt
- ¾ teaspoon ground black pepper
- 1 cup water
- 4 eggs

Directions:

1. Take a silicone tray, grease it with avocado oil and then crack the eggs into the cups of the tray. Switch on the instant pot, pour water in it, insert a trivet stand and place the silicone tray on it. Shut the instant pot with its lid in the sealed position, then press the 'manual' button, press '+/-' to set the cooking time to 7 minutes and cook at high-pressure setting; when the pressure builds in the pot, the cooking timer will start.
2. When the instant pot buzzes, press the 'keep warm' button, do a quick pressure release and open the lid. Ensure all eggs are cooked; egg whites should be firm, and yolk should be slightly jiggled.
3. Run a knife around each cup in the tray, then gently scoop out the egg and transfer to a serving plate. Season poached eggs with salt and black pepper and serve straight away.

Nutrition:

72 Calories

4.8g Fat

6.3g Protein

Spinach Egg Bites

Preparation Time: 5 minutes

Cooking Time: 20 minutes

Servings: 7

Ingredients:

- 4 Eggs
- ¾ cup Parmesan cheese, grated
- ¼ cup Heavy whipping cream
- ¼ cup Spinach, chopped
- ½ oz. Prosciutto, chopped
- ½ tsp. Ground black pepper
- 1/8 tsp. Salt
- 1 ½ cup Water

Directions:

1. Take an egg bite mold tray having seven cups and fill the cups evenly with prosciutto and spinach. Crack eggs in a bowl, add remaining ingredients except for water and whisk until smooth.
2. Switch on the instant pot, pour in water and place trivet stand in it. Pour egg mixture evenly over spinach and prosciutto, 4 tablespoons per cup or more until 3/4th filled, and then cover the pan with aluminum foil.
3. Place pan on the trivet stand, shut the instant pot with its lid in the sealed position, then press the 'manual' button, press '+/-' to set the cooking time to 10 minutes and cook at high-pressure setting; when the pressure builds in the pot, the cooking timer will start.
4. When the instant pot buzzes, press the 'keep warm' button, release pressure naturally for 10 minutes, then do a quick pressure release and open the lid.
5. Take out the tray, uncover it and turn over the pan onto a plate to take out the egg bites. Serve straight away.

Nutrition:

400 Calories

29g Fat

27g Protein

SNACK RECIPES

Keto Pancakes

Preparation Time: 5 minutes

Cooking Time: 15 minutes Servings: 4

Ingredients:

- 4 tsps. maple extract
- 8 eggs
- 4 tsps. cinnamon
- 8 coconut oil
- 2 ¾ pork rinds

Directions:

1. Put the pork in the blender and pulse until it becomes a fine powder. Then add the rest of the ingredients and mix them until smooth.
2. Heat a skillet to medium (300-400 ° F) and add the coconut oil into it. Pour batter into the skillet, fry it until golden brown (around 2 minutes), and of course, don't forget to flip it!
3. Bonus: If you want a sweet finish, you can add some fruit (for example strawberries) to it.

Nutrition:

2g Carbohydrates

43g Fat

24g Protein

510 Calories

Seafood Omelet

Preparation Time: 5 minutes

Cooking Time: 15 minutes

Servings: 4

Ingredients:

- 12 eggs
- 4 garlic cloves
- 1 cup mayonnaise
- 2 red chili peppers
- 10 oz. boiled shrimp or some seafood mix
- 4 tbsp. olive oil
- fennel seeds
- chives
- cumin
- 4 tbs. olive oil/butter
- salt and pepper to your taste

Directions:

1. Preheat your broiler Mix your seafood with olive oil, chili, cumin, minced garlic, salt, pepper, and fennel seeds. Then set it aside and cool to room temperature.
2. Add the chives (optional) and mayo to the cooled mixture. Whisk the eggs together, season them, and fry them in a skillet.
3. Add the mixture. When your omelet is almost ready, fold it, lower the temperature a bit, and let it set completely. Serve it immediately for the best taste.

Nutrition:

4g Carbohydrates

83g Fat

27g Protein

872 Calories

Salad Sandwich

Preparation Time: 5-10 minutes

Cooking Time: 5-10 minutes

Servings: 1

Ingredients:

- 2 oz. lettuce
- ½ avocado
- 1 cherry tomato
- 1 oz. Edam cheese (or any cheese you prefer)
- ½ oz. butter

Directions:

1. Rinse and cut the lettuce. Then use it as the base of your sandwich.
2. Cover the leaves with butter and place the cheese, avocado, and tomato on top of it.

Nutrition:

4g Carbohydrates

43g Fat

4g Protein

419 Calories

Ground Beef and Creamy Cauliflower Made in a Skillet

Preparation Time: 5 minutes

Cooking Time: 25 minutes

Servings: 4

Ingredients:

- 2 cloves garlic chopped
- 4 jalapeno peppers
- 2 tbsps. ghee
- 1 tsp. ground cumin
- 1 tsp. fish sauce
- 1 tbsp. coconut amino
- ¼ cup toasted sunflower seed oil
- ½ cup water
- 1 head cauliflower (400-500 g)
- 400-500 g lean ground beef
- 1 onion
- Salt
- Pepper
- 4 tbsps. mayo
- 4 eggs
- 1 tbsp. apple cider vinegar
- 1 tbsp. fresh parsley
- ½ ripe avocado

Directions:

1. Melt the ghee in your skillet around 300-400 °F (Medium-high). Add the onion, garlic, and jalapeno when your skillet is hot and cook it until softened (around 2-3 minutes).
2. Add the perfectly seasoned beef, keep cooking until the beef becomes completely brown. Then lower the heat to 200-300 °F (Medium-low), add the cauliflower, and continue for 2-3 minutes.
3. In the meantime, add the mayo (2 tbsp.), water, sunflower seed oil, coconut aminos, fish sauce, and cumin to a large cup and whisk them until thoroughly combined.
4. Pour it into the mixture in the skillet and stir it. Keep cooking for 3-5 minutes, until the liquids are absorbed. Remove it from the heat and spread it evenly into 4 plates.
5. Now you can crack the 4 eggs into your skillet and cook them to your liking. Meanwhile, mix 2 tbsp. mayo with your vinegar. Drizzle it over the skillet then garnish it with the diced avocado and the chopped parsley.

Nutrition:

688 Calories

52g Fat

38g Protein

14g Carbohydrates

Avocado and Salmon

Preparation Time: 5 minutes

Cooking Time: 15 minutes

Servings: 1

Ingredients:

- 1 ripe avocado
- salt
- 1 lemon (the juice part of it)
- 1 oz. goat cheese
- 2 oz. smoked salmon

Directions:

1. Cut the avocado, and remove the seed.
2. Mix the remaining ingredients, until they fuse well.
3. Place the cream inside the avocado.

Nutrition:

471 Calories

41g Fat

4g Carbs

19g Protein

Bacon and Eggs

Preparation Time: 5 minutes

Cooking Time: 10 minutes

Servings: 4

Ingredients:

- 8 eggs
- 5 oz. bacon, sliced optimal
- optional: tomatoes, parsley

Directions:

1. Cook bacon until crispy and then put it aside. Use the same pan for the eggs. Heat over a bit of medium heat and crack the eggs.
2. Cook them to your liking: repeat the same with the tomatoes, and parsley, if you are using them. Season them for your taste.
3. Tip: If you prefer it this way, you can fry the bacon and the eggs together to merge the flavors a bit better.

Nutrition:

1g Carbohydrates

22g Fat

15g Protein

272 Calories

Boiled Eggs with Mayo

Preparation Time: 3 minutes

Cooking Time: 10 minutes Servings: 4

Ingredients:

- 8 eggs
- 8 tbsps. mayo
- avocado (optional, but recommended)

Directions:

1. Boil water in a pot, and carefully put the eggs in the water. Boil the eggs: 5-6 minutes for soft, 6-8 minutes for medium, 8-10 minutes for hard-boiled eggs.
2. Tip: Serve it with simple with the mayonnaise and the avocado. Another option is to mash together the mayo and avocado. Or you can mix everything, by smashing the eggs into the mixture and creating a delicious cream.

Nutrition:

1g Carbohydrates

29g Fat

11g Protein

316 Calories

Breakfast Salad

Preparation Time: 3 minutes

Cooking Time: 15 minutes

Servings: 2

Ingredients:

- 2 eggs
- 2 oz. avocado (sliced preferred)
- 10 grape tomatoes
- 4 strips bacon
- black pepper, salt
- 1 tsp. red wine vinegar
- 2 tsps. virgin olive oil
- 3 shredded Lacinto kale

Directions:

1. Use a bowl to combine the kale, olive oil, vinegar, and a bit of salt, and smash them with your hands until the kale softens a bit. Cook the eggs to suit your style, medium boiled recommended here, and cook your bacon.
2. Divide it into 2 plates, use your toppings, the bacon, tomatoes, and avocado for your desired outcome, and don't forget to season it.

Nutrition:

292 Calories

18g Fats

13g Carbohydrates

18g Protein

Avocado Egg Wrapped into Prosciutto

Preparation Time: 5 minutes

Cooking Time: 10 minutes

Servings: 1

Ingredients:

- 2 eggs
- 2 avocados
- 2 tsps. olive oil
- salt and pepper
- 6 prosciutto slices
- chopped parsley, tomato slices to garnish

Directions:

1. Cook the eggs for hard-boiled eggs. Cut the avocados in half, then fill the middle of them with the eggs, and cut them half too.
2. Then wrap them inside the prosciutto, and fry it over medium heat in the olive oil for roughly 10 minutes. Cook until the bacon is crispy. Drain the excess oil before serving. Use the parsley and tomato slices to make it look even better.

Nutrition:

457 Calories

33g Protein

3g Carbs

54g Fats

Cauli Fritters

Preparation Time: 10 minutes

Cooking Time: 15 minutes

Servings: 2

Ingredients:

- 2 eggs
- 1 head of cauliflower
- 1 tbsp. yeast
- sea salt, black pepper
- 1-2 tbsp. ghee
- 1 tbsp. turmeric
- 2/3 cup almond flour

Directions:

1. Put cauliflower in your pot and boil it for 8 minutes. Add the florets into a food processor and pulse them. Add the eggs, almond flour, yeast, turmeric, salt and pepper to a mixing bowl. Stir well. Form into patties.
2. Heat your ghee to medium in a skillet. Form your fritters and cook until golden on each side (3-4 minutes). Serve it while hot.

Nutrition:

238 Calories

23g Fat

5g Carbohydrates

6g Protein

LUNCH RECIPES

Bacon Burger Stir Fry

Preparation Time: 10 minutes

Cooking Time: 20 minutes

Servings: 10

Ingredients:

- 1 lb. Ground beef
- 1 lb. Bacon
- 1 Small onion
- 3 garlic cloves
- 1 Cabbage

Directions:

1. Dice the bacon and onion. Mix the beef and bacon in a wok.
2. Mince the onion and garlic. Toss both into the hot grease. Slice and toss in the cabbage and stir-fry. Mix in the meat and season.

Nutrition:

32g Protein

22g Fats

357 Calories

Bacon Cheeseburger

Preparation Time: 15 minutes

Cooking Time: 30 minutes Servings: 12

Ingredients:

- 16 oz. Low-sodium bacon
- 3 lb. Ground beef
- 2 Eggs
- ½ Medium onion
- 8 oz. cheddar cheese

Directions:

1. Fry the bacon and chop it to bits. Shred the cheese and dice the onion. Mix the mixture with the beef and whisked eggs.
2. Grill 24 burger if desired.

Nutrition:

27g Protein 41g Fats 489 Calories

Cauliflower Mac & Cheese

Preparation Time: 15 minutes

Cooking Time: 20 minutes

Servings: 4

Ingredients:

- 1 Cauliflower
- 3 tbsp. Butter
- ¼ cup unsweetened almond milk
- ¼ cup Heavy cream
- 1 cup Cheddar cheese

Directions:

1. Slice the cauliflower into small florets. Shred the cheese. Prepare the oven to 450° Fahrenheit. Wrap baking pan with foil.
2. Melt two tablespoons butter. Mix the florets, butter, salt, and pepper. Roast the cauliflower on the baking pan for 15 minutes.
3. Warm rest of the butter, milk, heavy cream, and cheese in the microwave. Pour the cheese and serve.

Nutrition:

11g Protein

23g Fats

294 Calories

Mushroom & Cauliflower Risotto

Preparation Time: 5 minutes

Cooking Time: 10 minutes

Servings: 4

Ingredients:

- 1 cauliflower
- 1 cup Vegetable stock
- 9 oz. mushrooms
- 2 tbsp. Butter
- 1 cup Coconut cream

Directions:

1. Pour the stock in a saucepan. Boil and set aside. Prepare a skillet with butter and sauté the mushrooms.
2. Grate and stir in the cauliflower and stock. Simmer and add the cream. Serve.

Nutrition:

1g Protein

17g Fats

186 Calories

Pita Pizza

Preparation Time: 15 minutes

Cooking Time: 10 minutes

Servings: 2

Ingredients:

- ½ cup Marinara sauce
- 1 Low-carb pita
- 2 oz. Cheddar cheese
- 14 Pepperoni slices
- 1 oz. Roasted red peppers

Directions:

1. Set oven to 450° Fahrenheit.
2. Slice the pita in half and place onto a foil-lined baking tray. Rub with a bit of oil and toast for 2 minutes.
3. Pour the sauce over the bread. Sprinkle using the cheese and other toppings. Bake for 5 minutes.

Nutrition:

13g Protein

19g Fats

250 Calories

Skillet Cabbage Tacos

Preparation Time: 10 minutes

Cooking Time: 15 minutes

Servings: 4

Ingredients:

- 1 lb. Ground beef
- ½ cup Salsa
- 2 cups cabbage
- 2 tsp. Chili powder
- ¾ cup cheese

Directions:

1. Brown the beef and drain the fat. Pour in the salsa, cabbage, and seasoning.
2. Cover and lower the heat. Simmer 12 minutes using the medium heat.
3. Once softened, remove it from the heat and mix in the cheese.
4. Top with green onions.

Nutrition:

30g Protein

21g Fats

325 Calories

Taco Casserole

Preparation Time: 10 minutes

Cooking Time: 20 minutes

Servings: 8

Ingredients:

- 2 lbs. Ground beef
- 2 tbsp. Taco seasoning
- 8 oz. cheddar cheese
- 1 cup Salsa
- 16 oz. Cottage cheese

Directions:

1. Heat the oven to 400° Fahrenheit.
2. Combine the taco seasoning and ground meat in a casserole dish. Bake for 20 minutes.
3. Combine the salsa and both kinds of cheese. Set aside.
4. Drain away the cooking juices from the meat.
5. Mash the meat into small pieces.
6. Sprinkle with cheese. Bake in the oven for 20.

Nutrition:

45g Protein

18g Fats

367 Calories

Creamy Chicken Salad

Preparation Time: 10 minutes

Cooking Time: 30 minutes

Servings: 4

Ingredients:

- 1 lb. Chicken Breast
- 2 Avocados
- 2 Garlic Cloves
- 3 tbsp. Lime Juice
- 1/3 cup Onion
- 1 Jalapeno Pepper
- 1 tbsp. Cilantro

Directions

1. Set oven to 400 F. Line cooking sheet with foil. Layer the chicken breast up with some olive oil before seasoning.
2. Situate onto cooking sheet and put into the oven for 20 minutes. Let it cool and shred. Combine everything into a bowl and mash the avocado. Season well!

Nutrition:

20g Fats

4g Carbohydrates

25g Protein

Spicy Keto Chicken Wings

Preparation Time: 20 Minutes

Cooking Time: 30 minutes

Servings: 4

Ingredients:

- 2 lbs. Chicken Wings
- 1 tsp. Cajun Spice
- 2 tsp. Smoked Paprika
- ½ tsp. Turmeric
- 2 tsp. Baking Powder

Directions:

1. Prep the stove to 400 F. Dry chicken wings with a paper towel.
2. Mix all of the seasonings along with the baking powder. Toss the chicken wings in and coat evenly. Put on a wire rack that is placed over your baking tray.
3. Cook for 30 minutes. Pull out from the oven and flip to bake the other side for 30 minutes.

4. Take it out and set aside. Serve.

Nutrition:

7g Fats

1g Carbohydrates

60g Proteins

Cilantro and Lime Creamed Chicken

Preparation Time: 10 Minutes

Cooking Time: 20 minutes

Servings: 4

Ingredients:

- 4 Chicken Breast
- 1 tsp. Red Pepper Flakes
- 1 tbsp. Cilantro
- 2 tbsp. Lime Juice
- 1 cup Chicken Broth
- ¼ cup Onion
- 1 tbsp. Olive Oil
- ½ cup Heavy Cream

Directions:

1. Preheat skillet and place it over a moderate temperature. Season the chicken breast. Throw into the skillet and cook for 8 minutes on each side. Set aside
2. Stir in the onion and cook them for a minute then mix in cilantro, pepper flakes, lime juice, and the chicken broth.
3. Boil for 10 minutes. Whisk in your heavy cream and add in the chicken to coat.

Nutrition:

20g Fats

6g Carbohydrates

30g Proteins

SALAD RECIPES

Lunch Caesar Salad

Preparation Time: 15 minutes

Cooking time: 10 minutes

Servings: 2

Ingredients:

- 1 avocado, pitted
- 1 chicken breast, grilled and shredded
- 1 cup bacon, cooked and crumbled
- 3 tbsp. creamy Caesar dressing
- Salt and ground black pepper to taste

Directions:

1. Peel and slice avocado. In medium bowl, combine bacon, chicken breast and avocado.
2. Add creamy Cesar dressing, stir well. Season with salt and pepper, stir. Serve.

Nutrition:

329 Calories

2.99g Carbs

22.9g Fat

17.8g Protein

Asian Side Salad

Preparation Time: 35 minutes

Cooking Time: 12 minutes

Servings: 4

Ingredients:

- 1 green onion
- 1 cucumber
- 2 tbsp. coconut oil
- 1 packet Asian noodles, cooked
- ¼ tsp. red pepper flakes
- 1 tbsp. sesame oil
- 1 tbsp. balsamic vinegar
- 1 tsp. sesame seeds
- Salt and ground black pepper to taste

Directions:

1. Chop onion. Slice cucumber thin. Preheat pan with oil on medium high heat. Add cooked noodles and close lid.
2. Fry noodles for 5 minutes until crispy. Transfer noodles to paper towels and drain grease.
3. Combine cucumber, pepper flakes, green onion, sesame oil, vinegar, sesame seeds, pepper, salt and noodles. Mix well. Put in refrigerator at least for 20-30 minutes. Serve.

Nutrition:

397 Calories 3.97g Carbs

33.7g Fat

1.98g Protein

Keto Egg Salad

Preparation Time: 15 minutes

Cooking Time: 10 minutes

Servings: 4

Ingredients:

- 6 oz. ham, chopped
- 5 eggs, boiled and chopped
- 1 tsp. garlic, minced
- ½ tsp. basil
- 1 tsp. oregano
- 1 tbsp. apple cider vinegar
- 1 tsp. kosher salt
- ½ cup cream cheese

Directions:

1. In medium bowl, combine chopped ham with chopped eggs, stir. In another bowl, mix together garlic, basil, oregano, vinegar, and salt. Stir the mixture till you get homogeneous consistency.
2. Whisk together spice mixture and cream cheese. Add cream cheese sauce to egg mixture and stir gently. Serve.

Nutrition:

341 Calories

26g Fat

22.1g Protein

Cobb Salad

Preparation Time: 20 minutes

Cooking Time: 27 minutes

Servings: 1

Ingredients:

- 1 tbsp. olive oil
- 4 oz. chicken breast
- 2 strips bacon
- 1 cup spinach, chopped roughly
- 1 large hard-boiled egg, peeled and chopped
- ¼ avocado, peeled and chopped
- ½ tsp. white vinegar

Directions:

1. Heat up pan on medium heat and add oil. Add chicken breast and bacon, cook until get desired crispiness.
2. Add spinach and egg, stir. Add avocado and mix well. Sprinkle with white vinegar and stir.
3. Serve.

Nutrition:

589Calories

47.8g Fat

42g Protein

Bacon and Zucchini Noodles Salad

Preparation Time: 15 minutes

Cooking Time: 10 minutes

Servings: 3

Ingredients:

- 32 oz. zucchini noodles
- 1 cup baby spinach
- 1/3 cup blue cheese, crumbled
- ½ cup bacon, cooked and crumbled
- 1/3 cup blue cheese dressing
- Ground black pepper to taste

Directions:

1. Combine zucchini noodles, spinach, blue cheese, and bacon, stir carefully.
2. Add black pepper and cheese dressing, toss to coat. Serve.

Nutrition:

198 Calories

13.9g Fat

9.95g Protein

Chicken Salad

Preparation Time: 15 minutes

Cooking Time: 10 minutes

Servings: 3

Ingredients:

- 1 celery stalk
- 2 tbsp. fresh parsley
- 1 green onion
- 5 oz. chicken breast, roasted and chopped
- 1 egg, hard-boiled, peeled and chopped
- Salt and ground black pepper to taste
- ½ tsp. garlic powder
- 1/3 cup mayonnaise
- 1 tsp. mustard
- ½ tbsp. dill relish

Directions:

1. Wash and chop celery, parsley and onion. Place celery, onion and parsley in blender or food processor and blend well. Remove this mass from food processor and set aside.
2. Place chicken in food processor and pulse well. Add chicken to onion mixture and stir. Add egg, pepper and salt, stir gently. Add garlic powder, mayonnaise, and mustard and dill relish, toss to coat.
3. Serve.

Nutrition:

279 Calories

22.9g Fat

11.9g Protein

Asparagus Salad

Preparation Time: 20 minutes

Cooking Time: 5 minutes

Servings: 5

Ingredients:

- 2 lbs. asparagus, cooked and halved
- 1 tbsp. butter, melted
- ½ tsp. garlic powder
- 1 tsp. sesame seeds
- 1 tbsp. coconut oil
- 1 tbsp. apple cider vinegar
- 1 tsp. dried basil
- 1 tsp. salt
- 4 oz. Parmesan cheese, grated

Directions:

1. In bowl, combine asparagus, butter and garlic powder. Stir well. Add sesame seeds, coconut oil, vinegar, basil and salt. Mix well.
2. Set salad aside to marinate. Serve salad with grated Parmesan cheese.

Nutrition:

133 Calories

8.85g Fat

10g Protein

Apple Salad

Preparation Time: 15 minutes

Cooking Time: 5 minutes

Servings: 4

Ingredients:

- 1 medium apple
- 2 oz. pecans
- 16 oz. broccoli florets
- 1 green onion
- 2 tsp. poppy seeds
- Salt and ground black pepper to taste
- ¼ cup sour cream
- ¼ cup mayonnaise
- ½ tsp. lemon juice
- 1 tsp. apple cider vinegar

Directions:

1. Core and grate apple. Chop pecans and broccoli florets. Dice green onion. In bowl, combine broccoli, apple, pecans, and green onion. Stir well.
2. Sprinkle with poppy seeds, black pepper and salt, stir carefully. In another bowl, whisk sour cream, mayonnaise, lemon juice and vinegar.
3. Add this mixture to salad and toss to coat. Serve.

Nutrition:

249 Calories

22.9g Fat

4.8g Protein

Bok Choy Salad

Preparation Time: 20 minutes

Cooking Time: 10 minutes

Servings: 6

Ingredients:

- 10 oz. bok choy, chopped roughly
- 2 tbsp. coconut oil
- 4 tbsp. chicken stock
- 1 tsp. basil
- 1 tsp. ground black pepper
- 1 white onion, peeled and sliced
- ¼ cup white mushrooms, marinated and chopped
- 1 lb. tofu, chopped
- 1 tsp. oregano
- 1 tsp. almond milk

Directions:

1. Heat up pan on medium heat. Add bok choy, 1 tablespoon of oil and chicken stock.
2. Season with basil and black pepper. Add onion and close lid.
3. Simmer vegetables for 5-6 minutes, stirring constantly. Transfer vegetables to bowl and add mushrooms. Pour 1 tablespoon of oil in pan and heat it up again.
4. Add chopped tofu and cook for 2 minutes. Transfer tofu to bowl with vegetables and sprinkle with oregano. Pour in almond milk and toss to coat.
5. Serve salad.

Nutrition:

130 Calories

4.67g Carbs

11g Fat

6.9g Protein

Halloumi Salad

Preparation Time: 15 minutes

Cooking Time: 12 minutes

Servings: 2

Ingredients:

- 3 oz. halloumi cheese, sliced
- 1 cucumber, sliced
- ½ cup baby arugula
- 5 cherry tomatoes, halved
- 1 oz. walnuts, chopped
- Salt and ground black pepper to taste
- ½ tsp. olive oil
- ¼ tsp. balsamic vinegar

Directions:

1. Preheat grill on medium high heat. Put halloumi cheese in grill and cook for 5 minutes per side.
2. In mixing bowl, combine cucumber, arugula, tomatoes, and walnuts. Place halloumi pieces on top.
3. Sprinkle with black pepper and salt. Drizzle oil and balsamic vinegar, toss to coat.
4. Serve.

Nutrition:

448 Calories

3.98g Carbs

42.8g Fat

22.3g Protein

DINNER RECIPES

Zucchini Lasagna with Meat Sauce

Preparation Time: 10 minutes

Cooking Time: 6 hours Servings: 6

Ingredients:

- 4 small zucchinis, ends cut off (you can sub two large zucchinis)
- 1 pound(500gr) cooked ground meat or chopped meatballs
- 1/2 cup of your favorite pasta sauce
- 8 oz. mozzarella cheese, freshly shredded (about 2 cups), divided
- 15oz (425gr) container of part-skim ricotta cheese
- 1/2 cup Parmesan cheese, freshly grated - 2 eggs
- 1 tablespoon dried parsley flakes
- 1 teaspoon salt
- 1/2 teaspoon cracked black pepper

Directions:

1. Thinly slice (unpeeled) zucchini length-wise into thin strips, like lasagna noodles. It's easier to do this with a mandolin, but a large knife works just fine. (It's OK if some are only a few inches long.) Create cheese filling by combining 1 cup mozzarella cheese, ricotta cheese, Parmesan cheese, eggs, parsley flakes, salt, and pepper.
2. Create a layer of zucchini at the bottom of your slow cooker. (It's OK if pieces overlap.)
3. Top zucchini with a rounded 1/2 cup of cheese filling, 1 cup meat, and 1-3 tablespoons sauce.
4. Continue layering zucchini, cheese, meat, and sauce until you only have enough zucchini left for top layer. (A 6-quart slow cooker will have 4-5 layers and a 4-quart slow cooker will have 6-8 layers.)
5. Before you add the top layer of zucchini, add whatever sauce, meat, and cheese you have left. Top with zucchini and remaining 1 cup of mozzarella cheese.
6. Cover, and cook on low for 6-8 hours. Turn off slow cooker and let rest for at least 30 minutes, so juices become more set.

Nutrition:

386 calories

23.4g fat

112g protein

Chinese Pulled Pork – Char Siu

Preparation Time: 10 minutes

Cooking Time: 7 hours

Servings: 6

Ingredients:

- 1 kg pork shoulder or loin
- 1 cup chicken broth homemade is best
- 4 tablespoons sugar free tomato sauce homemade is best
- 1 tablespoon tomato paste
- 2 tablespoons garlic paste
- 4 tablespoons soy sauce
- 5 drops liquid sweetener
- 2 teaspoons ginger paste
- 1 teaspoon smoked paprika

Directions:

1. Place the pork in the bottom of the slow cooker. Combine all remaining ingredients and pour over the pork, ensuring it gets underneath as well.
2. Cook on low for 7 hours. Shred the pork with a fork and stir through the sauce, cooking for a further 30 - 60 minutes until the sauce has thickened to your liking, or eat it straight away.

Nutrition:

392 calories

23g fat

31g protein

Garlic Butter Chicken with Cream Cheese Sauce

Preparation Time: 10 minutes

Cooking Time: 7 hours

Servings: 6

Ingredients:

For the garlic chicken:

- 2- 2.5 lbs. of chicken breasts
- 1 stick of butter
- 8 garlic cloves, sliced in half to release flavor
- 1 tsp. salt
- Optional (but recommended) – 1 sliced onion or 2 tsps. of onion powder

For the cream cheese sauce:

- 8 oz. of cream cheese
- 1 cup of chicken stock
- salt to taste

Directions:

1. for the garlic chicken:
2. Place the chicken (thawed) in the slow cooker. Add the butter to the slow cooker. Place the garlic in the slow cooker, dispersing it around so it's not all in one spot. Sprinkle with salt. Cook on low for 6 hours. Remove and place on serving platter
3. for the cream cheese sauce:
4. In a pan, put the cup of chicken stock (or liquid from the slow cooker). Add the cream cheese and salt. Cook over medium-low heat until the sauce is combined and creamy. Pour over chicken.

Nutrition:

664 calories 38g fat 63g protein

Spinach Artichoke Chicken

Preparation Time: 10 minutes

Cooking Time: 4 hours Servings: 6

Ingredients:

- 16 oz. Cream Cheese softened
- 9 oz. Frozen Spinach cooked and drained well
- 14.5 oz. Artichoke Hearts chopped
- 1 tbsp. garlic
- 2 cups shredded mozzarella
- 3 lbs. boneless, skinless chicken (we used thighs)
- salt and pepper to taste

Directions:

1. Place the chicken in the bottom of a slow cooker. Salt and Pepper well. In a bowl, mix together cream cheese, spinach, artichokes, garlic and season with salt and pepper.

2. Stir in mozzarella cheese. Cook on low for 4-5 hours.

Nutrition:

460 calories

10g fat

34g protein

Midget
Momma

Bacon Wrapped Pork Loin

Preparation Time: 10 minutes

Cooking Time: 7 hours

Servings: 4

Ingredients:

- 2 lb. pork loin roast
- 4 strips uncooked bacon
- 1 package dry onion soup mix
- 1/4 cup water

Directions:

1. Rub pork loin with the dry onion soup mix. Pour any leftover into the bottom of the crock pot (any that fell off on my cutting board I scraped into mine).
2. Wrap the bacon around the roast and place into the crock pot. Pour in the water. Cook on High for 5 hours or Low for 7.

Nutrition:

388 calories

19g fat

40g protein

Chicken Cordon Bleu with Cauliflower

Preparation Time: 10 minutes

Cooking Time: 45 minutes

Servings: 4

Ingredients:

- 4 boneless chicken breast halves (about 12 ounces)
- 4 slices deli ham
- 4 slices Swiss cheese
- 1 large egg, whisked well
- 2 ounces pork rinds
- ¼ cup almond flour
- ¼ cup grated parmesan cheese
- ½ teaspoon garlic powder
- Salt and pepper
- 2 cups cauliflower florets

Directions:

1. Preheat the oven to 350 ° F and add a foil on a baking sheet. Sandwich the breast half of the chicken between parchment parts and pound flat. Spread the bits out and cover with ham and cheese sliced over.
2. Roll the chicken over the fillings and then dip into the beaten egg. In a food processor, mix the pork rinds, almond flour, parmesan, garlic powder, salt and pepper, and pulse into fine crumbs.
3. Roll the rolls of chicken in the mixture of pork rind then put them on the baking sheet. Throw the cauliflower into the baking sheet with the melted butter and fold. Bake for 45 minutes until the chicken is fully cooked.

Nutrition:

420 Calories

23g Fats

7g Protein

Sesame-Crusted Tuna with Green Beans

Preparation Time: 15 minutes

Cooking Time: 5 minutes

Servings: 4

Ingredients:

- ¼ cup white sesame seeds
- ¼ cup black sesame seeds
- 4 (6-ounce) Ahi tuna steaks
- Salt and pepper
- 1 tablespoon olive oil
- 1 tablespoon coconut oil
- 2 cups green beans

Directions:

1. In a shallow dish, mix the two kinds of sesame seeds. Season the tuna with pepper and salt. Dredge the tuna in a mixture of sesame seeds. Heat up to high heat the olive oil in a skillet, then add the tuna.
2. Cook for 1 to 2 minutes until it turns seared, then sear on the other side. Remove the tuna from the skillet, and let the tuna rest while using the coconut oil to heat the skillet. Fry the green beans in the oil for 5 minutes then use sliced tuna to eat.

Nutrition:

370 Calories

23g Fats

7g Protein

Rosemary Roasted Pork with Cauliflower

Preparation Time: 10 minutes

Cooking Time: 20 minutes Servings: 4

Ingredients:

- 1 ½ pounds boneless pork tenderloin
- 1 tablespoon coconut oil
- 1 tablespoon fresh chopped rosemary
- Salt and pepper
- 1 tablespoon olive oil
- 2 cups cauliflower florets

Directions:

1. Rub the coconut oil into the pork, then season with the rosemary, salt, and pepper. Heat up the olive oil over medium to high heat in a large skillet.
2. Add the pork on each side and cook until browned for 2 to 3 minutes. Sprinkle the cauliflower over the pork in the skillet.
3. Reduce heat to low, then cover the skillet and cook until the pork is cooked through for 8 to 10 minutes. Slice the pork with cauliflower and eat.

Nutrition:

320 Calories

37g Fats

3g Protein:

Grilled Salmon and Zucchini with Mango Sauce

Preparation Time: 5 minutes

Cooking Time: 10 minutes

Servings: 4

Ingredients:

- 4 (6-ounce) boneless salmon fillets
- 1 tablespoon olive oil
- Salt and pepper
- 1 large zucchini, sliced in coins
- 2 tablespoons fresh lemon juice
- ½ cup chopped mango
- ¼ cup fresh chopped cilantro
- 1 teaspoon lemon zest
- ½ cup canned coconut milk

Directions:

1. Preheat a grill pan to heat, and sprinkle with cooking spray liberally. Brush with olive oil to the salmon and season with salt and pepper.
2. Apply lemon juice to the zucchini, and season with salt and pepper. Put the zucchini and salmon fillets on the grill pan.
3. Cook for 5 minutes then turn all over and cook for another 5 minutes.
4. Combine the remaining ingredients in a blender and combine to create a sauce. Serve the side-drizzled salmon filets with mango sauce and zucchini.

Nutrition:

350 Calories 23g Fats 7g Protein

6g Carbohydrates

yummyhealthyeasy.com

Beef and Broccoli Stir-Fry

Preparation Time: 20 minutes

Cooking Time: 15 minutes Servings: 4

Ingredients:

- ¼ cup soy sauce
- 1 tablespoon sesame oil
- 1 teaspoon garlic chili paste
- 1-pound beef sirloin
- 2 tablespoons almond flour
- 2 tablespoons coconut oil
- 2 cups chopped broccoli florets
- 1 tablespoon grated ginger
- 3 cloves garlic, minced

Directions:

1. In a small bowl, whisk the soy sauce, sesame oil, and chili paste together. In a plastic freezer bag, slice the beef and mix with the almond flour. Pour in the sauce and toss to coat for 20 minutes, then let rest.
2. Heat up the oil over medium to high heat in a large skillet. In the pan, add the beef and sauce and cook until the beef is browned.
3. Move the beef to the skillet sides, then add the broccoli, ginger, and garlic. Sauté until tender-crisp broccoli, then throw it all together and serve hot.

Nutrition:

350 Calories 19g Fats

37g Protein

6g Carbohydrates

Conclusion

If you're eyeing for a diet that works and gives you the results you want, then it's time to take your health and performance to the next level. It is also one of the most effective ways to reduce appetite and feel full. It's also a natural heal for diabetes, epilepsy, and Alzheimer's disease.

This keto diet is a low-carbohydrate, high-fat diet that increases your body's ability to burn fat as fuel.

The ketogenic diet main purpose is to cause your body to make ketones, which are compounds produced by the liver used as an alternative fuel source for your body instead of glucose (sugar). These ketones then serve as a fuel source all over the body, especially for the brain.

In less than 5 years, the keto diet has gone from a notorious fad diet to a well-respected high-protein health and wellness regimen. An increasing amount of people are deciding on living without carbohydrates. They rely solely on fat-forming foods like meat, fish, eggs, cheese, butter, and coconut oil for their caloric intake. This trend has been gaining ground for more than 30 years with people following such diets as Atkins. The reason for this recent spike in popularity can be attributed partially to the 2014 documentary The Carbohydrate Addict, which focuses on Dr. Robert Atkins' theory that carbohydrates play a central role in heart disease.

Not long after its release, the ketogenic diet was used as the backbone of a new trend known as "ketogenic" or "low carb" diets. These low-carb diets claim that by restricting carbohydrates from your daily meal plan, your body will become efficient at burning fat as fuel instead of glucose. The purpose of following such a regimen is to create your own "ketogenic" state wherein your body will naturally become efficient at burning fat stored within your liver and muscles for energy instead of carbohydrates.

The first thing people should be prepared for is the signs of the body-switching over to ketosis. These include bad breath, weight loss, appetite decrease, and potential weakness in the beginning stages. It is normal to have these reactions while doing the keto diet. It can also be helpful to be familiar with the keto flu's signs and symptoms, which can affect people at varying severities. Finally, they should have an idea of how long they will need to stay on a diet to achieve their desired results. Some people choose to do standard keto until they reach their weight loss goals and then choose a less vigorous form of the diet to keep the pounds off.

For people who have stomach issues when starting the diet, switching to fats that are easier to digest can be a smart move for the beginning stages. Adding fiber to the diet can also help regulate the gut and ease those uncomfortable symptoms.

After making it through the keto flu, here are the benefits of the diet. Some people decide to stay on the meal plan long term. Although it is not recommended to do full keto for longer than a year, keeping some form of the diet long-term can help to ensure the goals met are not lost. To ensure that staying on a diet is simple and easy, people should focus on eating quality fats that smoothly help their brain and body function. If the body doesn't have to work hard to digest food, the person will usually have more energy and feel better overall.

CPSIA information can be obtained
at www.ICGtesting.com
Printed in the USA
LVHW020905270521
688665LV00014B/664